THE AMAZING HUMAN BODY

THE HUMAN
SKELETAL SYSTEM

by Tammy Gagne

BrightPoint Press

San Diego, CA

BrightPoint Press

For more information, contact:
BrightPoint Press
PO Box 27779
San Diego, CA 92198
www.BrightPointPress.com

LIBRARY OF CONGRESS CATALOGING-IN-PUBLICATION DATA

Name: Gagne, Tammy, author.
Title: The human skeletal system / by Tammy Gagne.
Description: San Diego, CA: BrightPoint Press, 2025 | Series: The amazing human body | Audience: Grade 7 to 9 | Includes bibliographical references and index.
Identifiers: ISBN: 9781678209629 (hardcover) | ISBN: 9781678209636 (eBook)
The complete Library of Congress record is available at www.loc.gov.

CONTENTS

AT A GLANCE

- The skeletal system is a framework of bones that gives the body its shape and makes movement possible.

- Most people have between 206 and 213 bones. Bones contain calcium, collagen, and phosphorus.

- The skeletal system can be divided into two parts. They are the axial skeleton and the appendicular skeleton.

- The skeletal system helps other body systems function. These include the muscular system and the circulatory system.

- Common illnesses that affect the skeletal system include osteoarthritis, bursitis, and osteoporosis. These conditions are most common in older adults.

- Fracturing, or breaking, bones can be painful. But broken bones usually heal over time.
- The human body needs calcium, vitamin D, and protein for bones to grow and stay healthy.
- People can protect their bones from injuries by wearing seat belts, using helmets and padding when taking part in recreational activities, and exercising safely.

A BAD BREAK

It was March 17, 2024. College basketball fans filled Bridgestone Arena in Nashville, Tennessee. The NCAA Southeastern Conference Tournament final had just begun. The Florida Gators were playing the Auburn Tigers. Florida center Micah Handlogten was off to a great start. He had two rebounds only 2 minutes into the game.

He went for a third rebound. But he came down the wrong way. He hit the

Leg injuries are common in basketball players. This is partly because the sport demands frequent jumping.

floor hard. Some people in the crowd heard a snapping sound coming from his left leg. The player's parents stood nearby as medical workers rushed to help their son. The talented center was clearly unable to finish the game. He headed for Vanderbilt University Medical Center.

Two days later, Handlogten was supporting his team from the sidelines. He stood with the help of crutches. He'd had

Crutches help people with leg injuries support themselves without putting weight on injured limbs.

surgery on his broken leg. "It's a long recovery, but I'm looking forward to it," he said. "I'm going to come back better and stronger for it."[1]

BREAKING AND HEALING

Many people admired Handlogten's positive attitude. It had been obvious that the break was a painful one. As soon as it happened, Handlogten rolled onto his side. He covered his face with his hands in agony.

But his skeletal system soon began to heal itself. The skeletal system is the body's framework of bones. It gives the body its shape and makes movement possible. It also protects many of the body's most important organs. And when a bone breaks, it starts healing almost immediately.

WHAT IS THE SKELETAL SYSTEM?

The human skeletal system is a group of bones and tissues within a person's body. The parts work together to do many important things. The skeletal system supports and protects the body. It also produces most of the body's blood cells.

The body's soft internal organs rely on the hard bones of the skeletal system for protection. The skull is made up of several bones in the head. The skull protects

The frontal bone is the bone in the skull behind a person's forehead. This bone protects the brain.

MAJOR STRUCTURES OF THE HUMAN SKELETAL SYSTEM

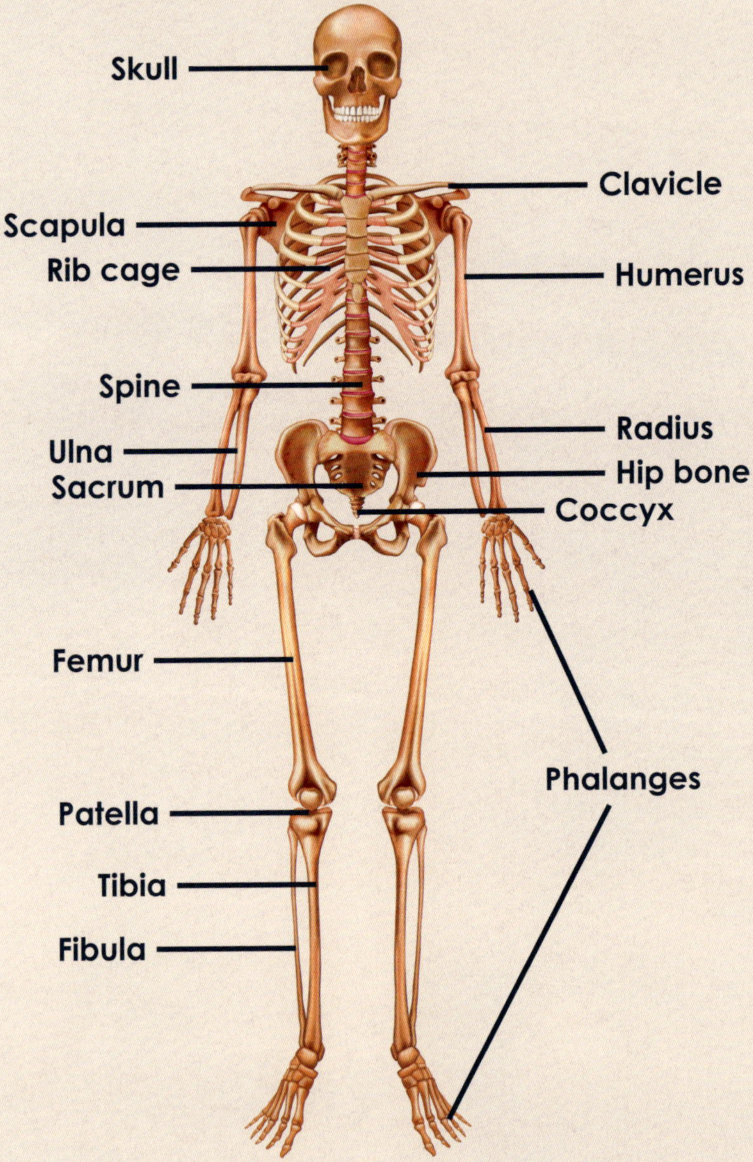

This diagram displays the major bones and bone structures of the human skeletal system.

the brain. The spine extends from the base of the skull to the pelvis. It protects the spinal cord. The rib cage is a series of curved bones in the chest. These bones protect the heart and the lungs. Without bones, organs would move around inside the body and fail to work properly.

Bones do more than provide structure for the body. They also make blood cells. Blood cells form in the bone marrow. This soft tissue is found inside most bones. Blood cells begin as **stem cells**. They later grow into red blood cells, white blood cells, or platelets. Red blood cells carry oxygen from the lungs to the rest of the body. White blood cells help the body fight infections. And platelets make it possible for the blood to clot. A blood clot is a thick mass that forms in blood vessels. Clotting is how the body stops bleeding.

WHAT ARE BONES MADE OF?

Bones are made mostly of calcium and collagen. About 99 percent of the calcium in a human's body is found in the bones and teeth. Calcium is a mineral that makes bones hard and strong. Bones store calcium and other minerals for use by other body parts. For example, the body uses calcium for cell and nerve function.

Collagen is a protein that makes bones **flexible**. This increases their strength. Some medical experts compare the calcium in bones to concrete. Concrete is hard. But it can break under certain kinds of pressure. This is why builders place steel

Steel bars raise concrete's ability to withstand being stretched. Collagen does the same thing for bones.

bars inside concrete. The bars make the material stronger. Collagen works like these steel bars. It strengthens the calcium. This makes bones less likely to break.

Bones also contain phosphorus. About 85 percent of the phosphorus in the body is in the bones and teeth. The body uses this mineral in several ways. It helps the body store and use energy. It also aids in the growth and repair of body tissues.

Sodium is another mineral stored in the bones. It has many functions in the body. For example, it helps muscles work.

People's bones change over time. Bones can reshape and repair themselves as needed. Lots of pressure from repeated exercise can make part of a bone become thicker. And bones will usually heal by themselves after breaks. Science journalist

Babies are born with many small, soft bones. Their bones help them remain flexible inside the womb and during birth.

Amber Dance writes, "The skeleton is not just a nice set of mechanical supports. Bones constantly remodel themselves in response to the body's needs."[2]

THE BONES OF THE BODY

People usually have between 206 and 213 bones. Some people may have slightly more or fewer bones than others in certain parts of the body. These parts include the spine and the rib cage. Most babies have 270 to 300 bones when they are born. As children grow, some of their bones **fuse** together.

The femur is a bone in the leg. It is the largest bone in the human body. It runs from the hip to the knee. The average femur is about 18 inches (46 cm) long. It weighs 14 ounces (400 g). It is also the body's strongest bone. It can support up to thirty times a person's weight.

The human body's smallest bone is in the ear. It is called the stapes. It measures only 0.1 inches (2.5 mm) long. The stapes is shaped like a horseshoe. There are two other small bones in the ear. These are the malleus and the incus. These bones help carry sound waves deeper into the ear. Eventually, these waves become signals

Paired Bones

Most of the bones in the human body come in pairs. Of the 206 human bones, 172 are paired. For example, people have a femur in both of their legs. This pairing gives the body its symmetrical appearance.

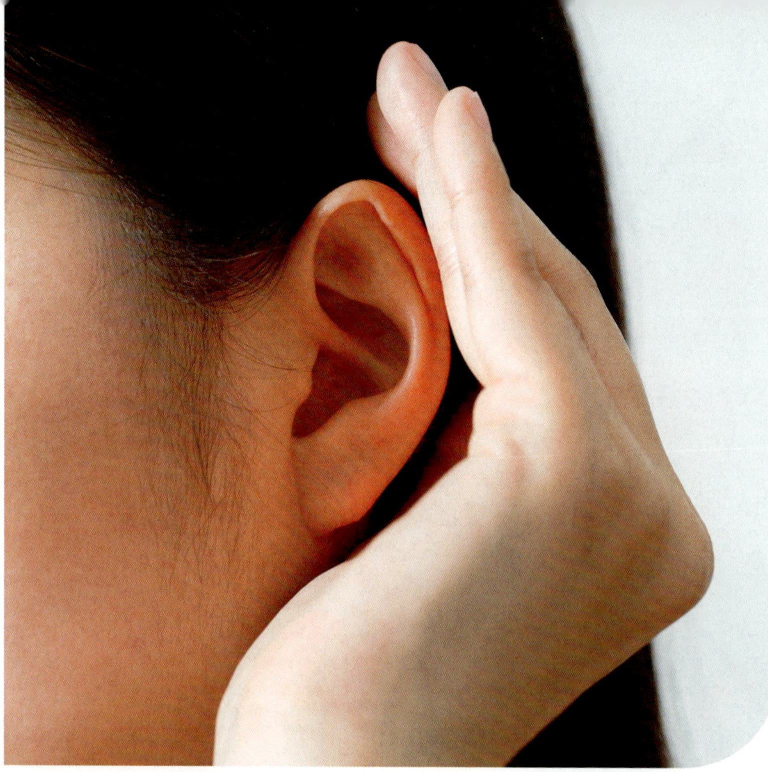

The bones in the ear increase the strength of sound waves by 100 times.

that the brain can understand. Without the ear bones, people could not hear properly.

About half of the bones in a person's body are in the hands and feet. Each hand contains twenty-seven bones. Each foot has twenty-six. These bones are called phalanges. They are arranged in similar ways in the hands and feet. Each thumb has two bones. Each finger has three. Like

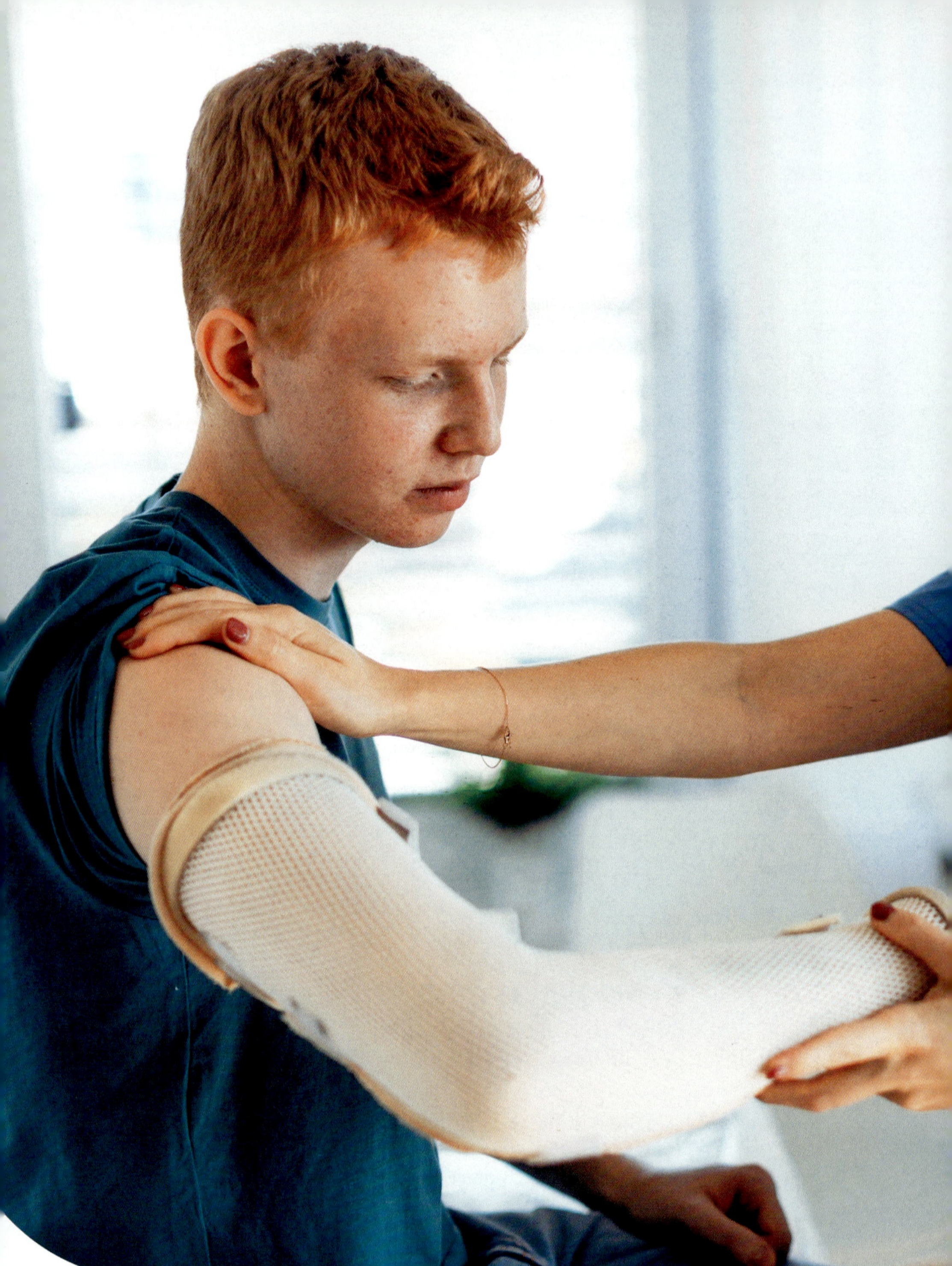

A cast can help a broken bone in the arm heal properly.

the thumb, the big toe has two bones. The other toes have three.

Steve E. Meadows is an orthopedic surgeon. This kind of doctor operates on bones. He writes, "Any bone in the body can be fractured or broken, but certain bones are more prone to these injuries."[3] About 50 percent of broken bones in adults occur in the arm. Children also break arm bones most often. Forty percent of breaks in children occur in the arm. Other common breaks happen in the ankle, collarbone, and wrist.

CHAPTER TWO

HOW DOES THE SKELETAL SYSTEM WORK?

The skeletal system is divided into two main sections. One is called the axial skeleton. It includes the skull, spine, and rib cage. This section supports and protects the organs that keep the body alive. These include the brain, spinal cord, heart, and lungs.

These organs are part of three other important systems of the body. The brain and the spinal cord are the primary parts

The bones of the spine are called vertebrae.

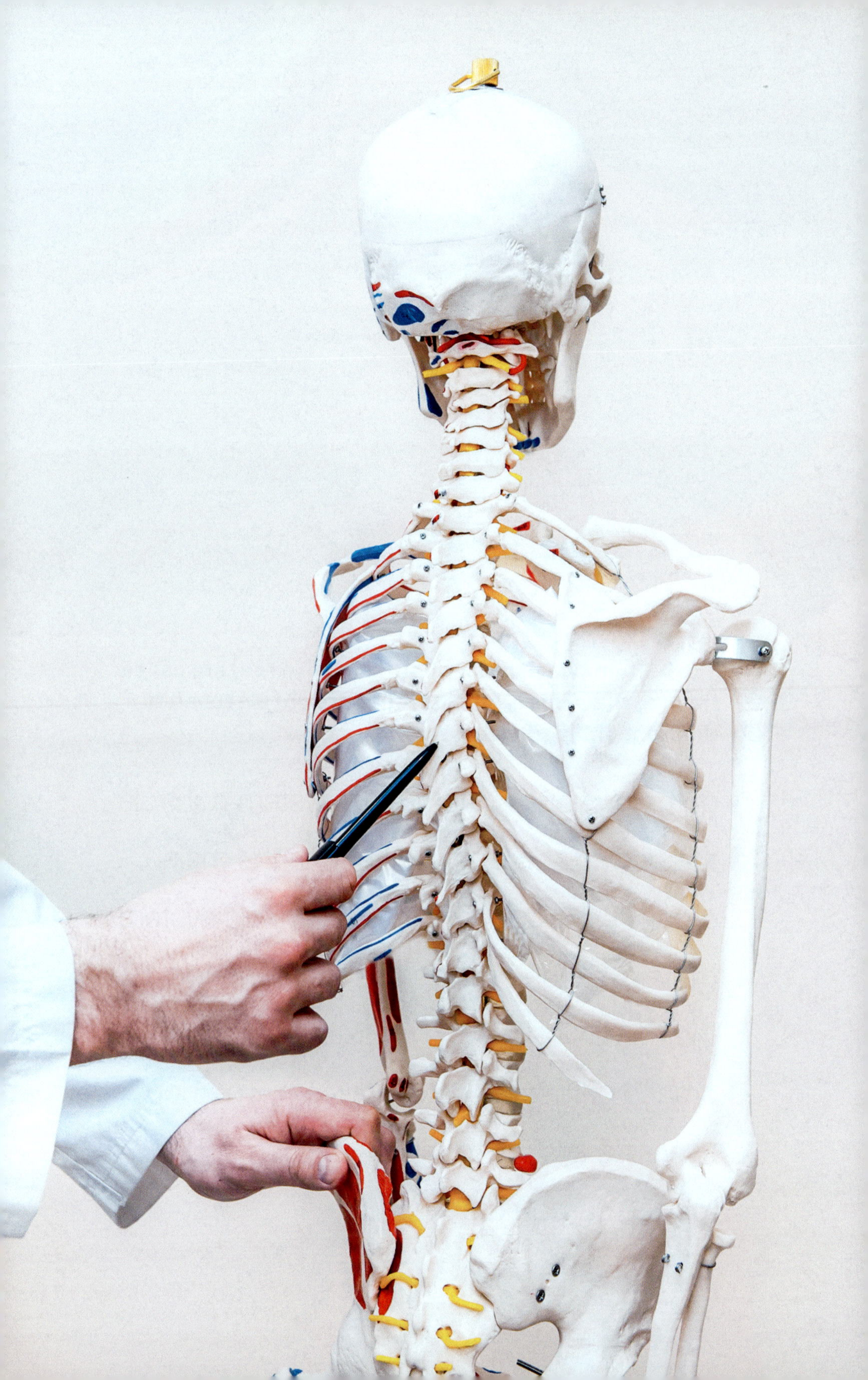

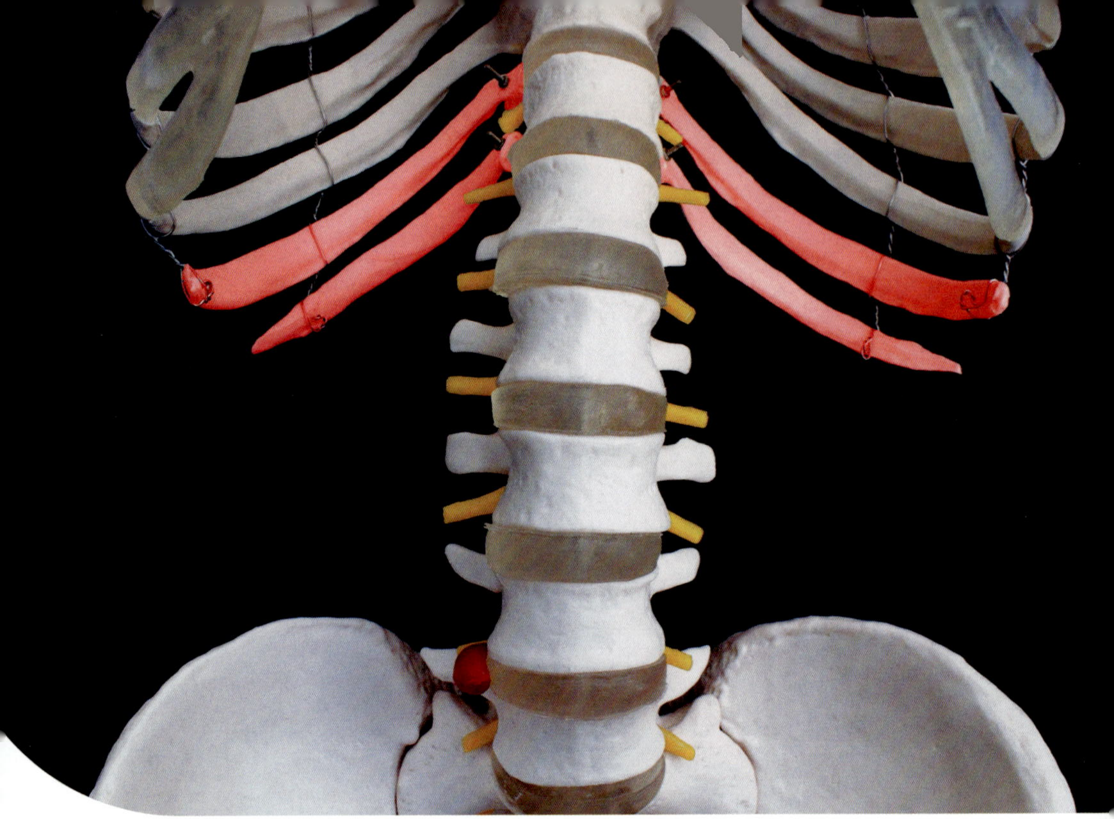

The bottom two pairs of ribs (modeled in red) are called floating ribs. This is because they are not attached to the front of the rib cage.

of the nervous system. This system controls the body's actions. The heart is the main organ of the circulatory system. This system delivers oxygen and nutrients throughout the human body. The lungs are the main organs of the respiratory system. The body uses this system to take in oxygen and release carbon dioxide.

People have eighty bones within their axial skeleton. Most form the spine and the rib cage. The spine is made up of twenty-four bones. They stack on top of one another. This forms a canal through which the spinal cord runs. These bones protect the spinal cord and allow it to be flexible. The rib cage also contains twenty-four bones. There are twelve ribs on each side of the body.

The Hyoid Bone

There is only one bone in the human body that isn't connected to another bone. It is called the hyoid bone. It is a U-shaped bone in the mouth. It is located at the base of the tongue. The hyoid bone helps with chewing, swallowing, and talking.

The axial skeleton plays a big role in posture, or the way a person stands and sits. Dana Santas helps train professional athletes. She explains, "It's the axial skeleton, [composed] of the head, spine and rib cage, that holds a human erect in what we call 'posture.'"[4]

THE APPENDICULAR SKELETON

The appendicular skeleton is made up of bones in and near the body's limbs. The shoulders, arms, and hands are all part of the appendicular skeleton. So are the hips, legs, and feet. When a person decides to move a body part, the brain sends a signal to the right muscles. The muscles are attached to the appendicular skeleton by tough cords called tendons. The muscles

Muscles that move bones come in pairs. Each muscle pulls the bone in a single direction. Raising and lowering a weight requires multiple pairs of muscles.

tighten to move the bones they are attached to. This moves the body part.

Much of the body's movement takes place at joints. Joints are areas where bones meet. The bones are connected to each other by special tissues. These include cartilage and ligaments. Cartilage is a firm, flexible tissue that pads bones. Ligaments are thick bands of tissue that hold bones together. Muscles pull bones at some joints to make many kinds of movements. These include bending and rotating. For example, the elbow joint allows people to push or pull things by bending the arm.

The human shoulder is made up of two main bones. The first is the clavicle, or collarbone. The second is the scapula,

The kneecap allows muscles in the thigh to bend the leg at the knee with greater strength.

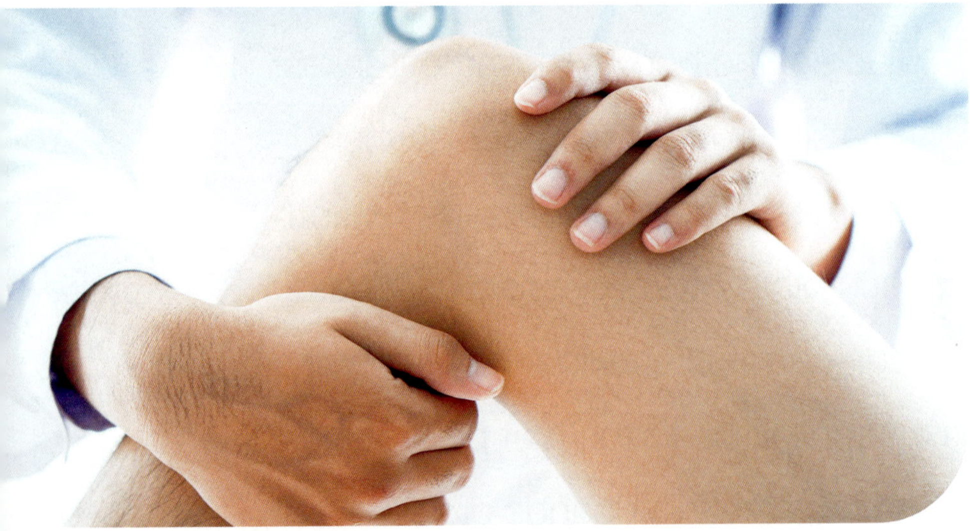

or shoulder blade. They allow people to move their shoulders and arms. A person's arms contain three bones. The humerus is in the upper arm. The radius and ulna are in the lower arm. The wrists and hands are also part of the appendicular skeleton.

Three bones in the lower appendicular skeleton connect to the axial skeleton. They are the hip bone, the sacrum, and the coccyx. These bones make up the pelvis. The pelvis also attaches to the lower limbs. The bones in the lower appendicular skeleton make movement such as walking or dancing possible. Like the arms, the legs each contain three main bones. The femur is in the upper leg. The tibia and fibula are in the lower leg. The patella, or kneecap, is between the upper and lower leg bones.

WORKING TOGETHER

Each system in the human body has its own purpose. But one system may also work with other systems. Movement is possible because the skeletal system works together with the muscular system. The body contains more than 650 muscles attached to bones. These muscles are called skeletal muscles.

The skeletal system also helps the nervous system work. Bones store calcium. The nervous system needs this mineral to work properly. The skeletal system also has special nerve cells in the joints. These cells send signals to the brain about the position of the body.

The skeletal system works with the circulatory system as well. Bone marrow

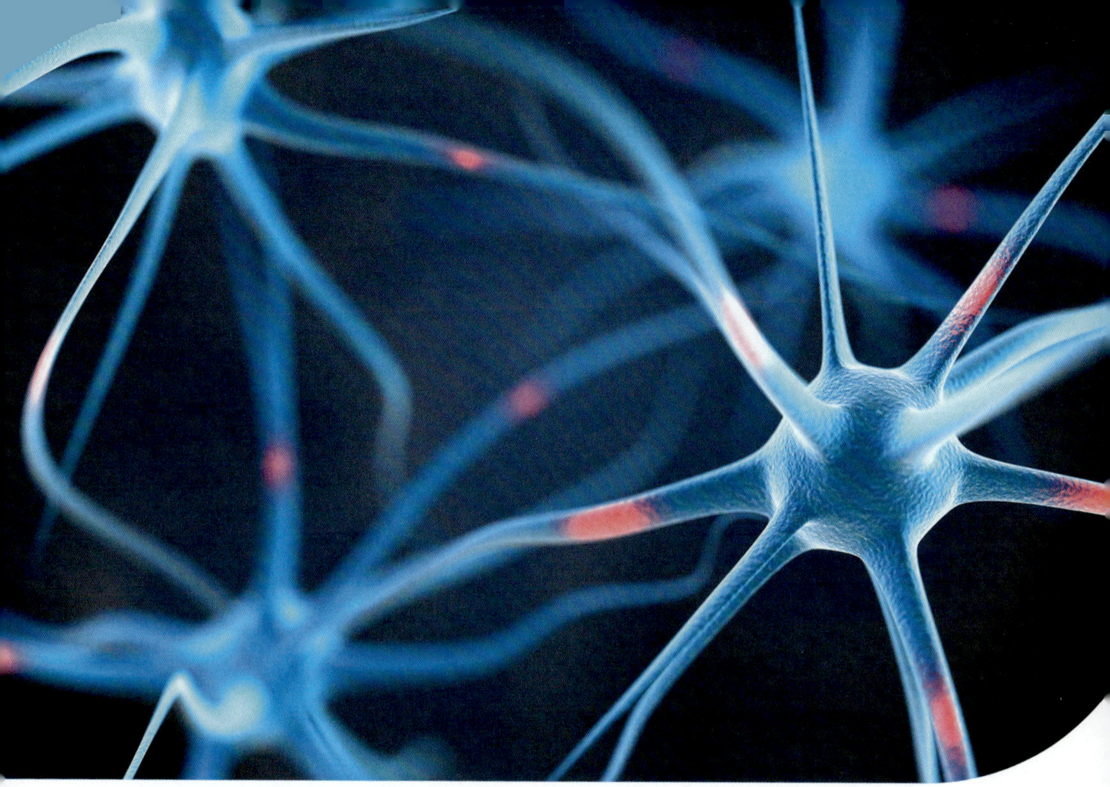

Cells in the nervous system use calcium to send electrical signals to each other.

produces the body's blood. As much as 10 percent of that blood returns to the bones. This blood helps bones grow and repair themselves.

CHAPTER THREE

WHAT CAN GO WRONG WITH THE SKELETAL SYSTEM?

Bones are very tough. But they can still break. People can break bones in accidents. These include car crashes and falls. Sometimes bones break into two or more pieces. This is called a complete fracture. Bones can also develop thin cracks. These are called hairline fractures. Broken bones can usually repair themselves. Orthopedic surgeon

Although bones usually heal themselves, treatment can make the healing process faster and safer.

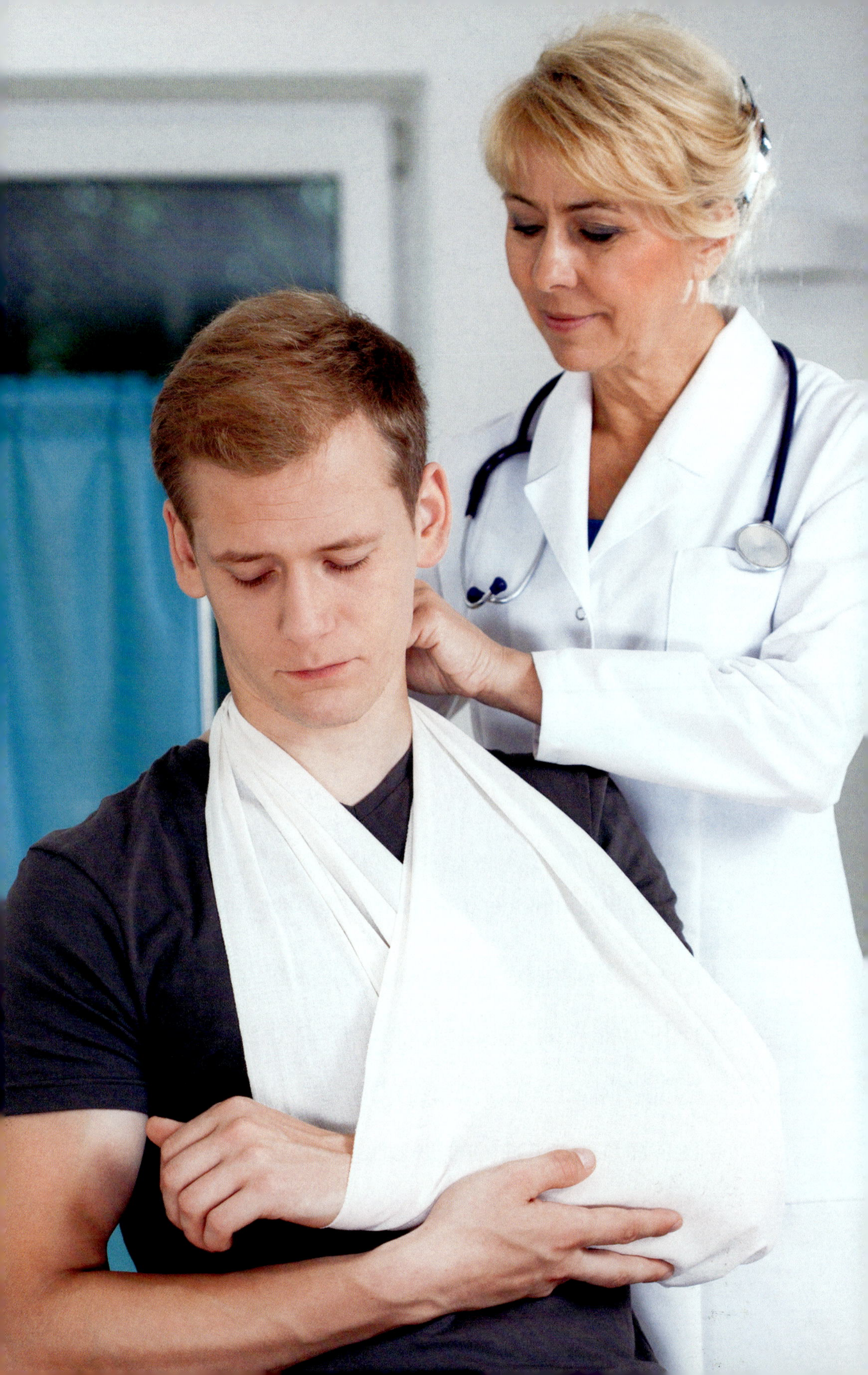

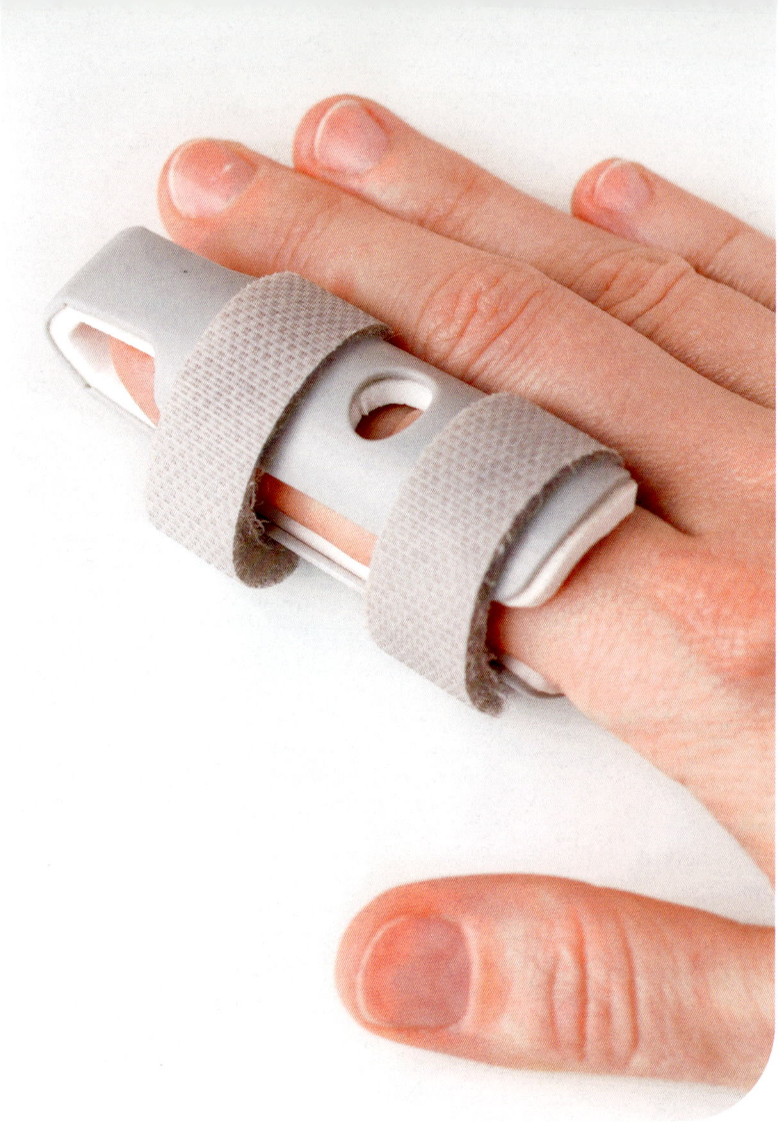

Splints are commonly used to treat broken fingers.

Rob S. Williams explains, "Broken bone ends will 'knit' back together in time."[5]

Breaking a bone can be very painful. The area around the break usually becomes red and swollen. Blood starts to gather around the break. The blood is replaced by a soft callus within about a week. This soft bone material is the first stage of healing. The callus hardens over the next several weeks. After about 6 weeks, the person can start using the body part normally again. However, some breaks can take longer to heal. Others may not fully heal on their own.

Casts and **splints** can hold broken bones in place while they heal. These devices are commonly used on broken arms and legs. But not all broken bones need casts or splints. Some bones often heal better without these objects. These include the collarbone and wrist. But people still need to avoid using a broken bone until it heals. Sometimes surgery is needed for a bone to heal properly.

OTHER SKELETAL SYSTEM INJURIES

People can dislocate bones. This happens when two bones separate at a joint. Dislocating a bone can be very painful. It can make it impossible to move the body part. Like breaks, dislocations can require surgery. But sometimes a doctor can guide a dislocated bone back into place without surgery. Shoulders and fingers are among the most commonly dislocated body parts.

Not all injuries to the skeletal system involve bones. Ligaments can be injured as well. Injuries to the ligaments around joints are called sprains. These injuries happen when ligaments are stretched too far or torn. Sprains can occur when a person lands too hard on a body part. They can

Twisted ankles are common injuries among athletes.

also result from twisting a joint. Dislocations can also injure ligaments. Minor sprains can be treated with rest and ice packs. Serious sprains may need surgery.

Sometimes parts of the skeletal system can pinch a nerve. This happens when tissues such as bones or tendons move or change shape. This can put pressure on a nearby nerve. A pinched nerve can be painful. It can also cause numbness and weakness. For example, a pinched nerve in the wrist can weaken a person's hand. The neck, back, and arms are other places where pinched nerves commonly occur. Pinched nerves can sometimes fix themselves with rest. Other cases may require surgery.

MISSING OR ABNORMAL LIMBS

Not all people have two arms and two legs. Some people are born missing one or more limbs. Other people are born with limbs that formed abnormally. For example, bones in the foot might be missing. This could cause the foot to develop in an abnormal shape. Missing or abnormal limbs may be the result of illness or **genetics**.

Some artificial legs are designed to be used by runners.

In some cases, people suffering from an illness or injury need an amputation. This surgery removes one or more limbs from a person's body.

An artificial limb called a prosthesis can help a person who is missing a limb move with more freedom. Physical therapy can

Doctors use X-ray scans to observe conditions affecting a patient's bones.

help a person relearn certain movements. Learning how to use a prosthesis and going through physical therapy can take time. But the process can have a positive effect on the lives of people missing limbs.

SKELETAL CONDITIONS

One of the most common conditions of the skeletal system is osteoarthritis. In 2024, journalist Vanessa Caceres reported, "There are 32.5 million adults in the [United States] living with osteoarthritis."[6] This condition affects the bones at the joints. Osteoarthritis happens as cartilage breaks down over time. This tissue acts as a cushion for the ends of bones. But movement becomes painful as the tissue breaks down. Osteoarthritis can affect any joint.

But it most often affects people's hands, knees, hips, and spine. Medications can treat osteoarthritis.

The skeletal system has small fluid-filled sacs that also act as cushions. They are called bursae. They can become irritated. This condition is called bursitis. Like osteoarthritis, bursitis affects a person's joints. This condition often affects the shoulders, elbows, or hips. Rest is the most common treatment for bursitis.

Osteoporosis is a condition that causes bones to become weak and **brittle**. When people with this condition fall, they are more likely to break a bone. Poor posture is one sign of osteoporosis. People with this condition often look hunched over. Women are more likely to suffer from osteoporosis

There is no one correct posture. But having a limited range of motion can be a sign of a medical condition.

than men. Osteoarthritis, bursitis, and osteoporosis are all more likely to occur as people age.

Scoliosis is another common bone condition. People with scoliosis have spines that curve abnormally. The spine may be shaped like the letters *S* or *C*.

Doctors do not know exactly what causes this condition. It affects 2 to 3 percent of people living in the United States. It most often develops in children between 10 and 15 years old. Some cases are not severe. They require no treatment. In other cases, the spine may be very curved. Growing children can wear braces to stop the curving. Surgery can correct the condition in serious cases.

Bone Cancers

Cancers that develop in the bones are called sarcomas. They are rare. Sarcomas make up less than 1 percent of all types of cancers diagnosed. Cancer can also occur in the bone marrow. This type of cancer is called multiple myeloma. It is usually found in older adults.

Treatments for bone cancer include surgery, anticancer medicine, and radiation therapy. Radiation therapy uses high-energy waves to kill cancer.

CHAPTER FOUR

WHAT CAN HUMANS DO TO KEEP THE SKELETAL SYSTEM HEALTHY?

Keeping bones healthy begins with giving them plenty of nutrients. Eating a healthy diet is the best way to do this. The human body needs calcium. But it does not make this nutrient. If the body doesn't get enough calcium through food, it will take it from the bones. The body cannot do this forever. *Harvard Health Letter* editor Heidi Godman compares this to taking lots of

One cup (237 mL) of milk contains almost a quarter of the calcium most people need per day.

money out of the bank. She says, "If you do it every day, you'll run out of money."[7]

People can make sure they get enough calcium by eating dairy foods. These include milk, cheese, and yogurt. The mineral is also found in nuts, seeds, beans, and soy. Some vegetables are also rich in calcium. They include leafy greens, artichokes, and squash. Fruits and seafood contain calcium as well.

Vitamin D is also important for bone health. This substance helps the body absorb calcium. Only a few foods contain vitamin D. These include salmon and certain mushrooms. Some companies add vitamin D to products that do not naturally contain it. These products include milk and orange juice.

Unlike calcium, vitamin D is made by the human body. Sunlight reacts with a chemical in the skin. Eventually, this process produces vitamin D. However, there is a risk of skin cancer when people spend too much time in the sun. This is why people should limit their sun exposure. They should also wear sunscreen.

Sunscreen reduces the harmful effects of the sun on skin. Using sunscreen does not harmfully lower vitamin D levels.

Bones also need protein to grow and repair themselves. This nutrient is found in many foods, including dairy, fish, poultry, and whole grains. Vegetables such as corn, broccoli, and asparagus also contain small amounts of protein.

To keep their bones healthy, people must get enough to eat. Being underweight weakens the bones. Over time, being

Nutrient Supplements

One way to make sure the body gets enough vitamin D is by taking supplements. These are products that provide nutrients. People who do not consume enough calcium in their diet can also take supplements. But it is best to get calcium from food. The body absorbs calcium from food better than it does calcium from supplements.

Smoking can have especially negative effects on bone health in people who begin smoking at a young age.

underweight can raise a person's chance of osteoporosis. Getting enough to eat is especially important for young people. This is because their bones are still growing. Having weak bones at a young age can affect bone health far into the future.

Another factor that raises the risk of osteoporosis is smoking. Tobacco contains nicotine. This chemical is addictive. It is also bad for the bones. It slows their growth.

Older people who smoke are up to 40 percent more likely to break their hips than people who do not smoke.

EXERCISING SAFELY

Staying active is one of the best ways to keep bones healthy. People who exercise

Weight-bearing exercise is good for bones. It also improves heart health.

regularly tend to have stronger bones than people who do not. This is because bones react to exercise by becoming denser over time. Weight-bearing exercise is especially good for bones. This is exercise that involves supporting one's own weight. Walking, hiking, and running are examples of this kind of exercise.

People who exercise also improve their balance and **coordination**. Improving these skills can reduce the risk of accidents that can damage bones. It is important to prepare the body for exercise. Warm-up exercises can help prevent injuries. "But the [warm-ups] and stretching that you undertake should be done safely and must include the correct **technique**," says Becki Andrus.[8] Andrus writes for Orthopedic

Associates in Dallas, Texas. Stretching too far or in the wrong way can cause an injury.

PROTECTING THE BONES

Everyday activities such as riding in a car or cycling can lead to accidents. People can reduce their chances of breaking bones in accidents by taking simple safety precautions. Wearing a seat belt is one of the easiest ways to prevent spine fractures in car accidents.

Helmets help prevent skull fractures in bicycle crashes. Wearing a helmet is also smart during activities such as rollerblading or riding a scooter. Other safety gear can prevent broken bones while taking part in these activities as well. This gear includes knee, elbow, and wrist pads. Shin guards

Wearing a seat belt in the front seat of a car can reduce the risk of moderate to critical injury by 50 percent.

Wrist guards can help protect a skateboarder's wrists from breaks in the event of a fall.

and other padding can protect bones while people play sports such as soccer.

The human skeletal system makes amazing things possible. It helps people run and climb. It allows them to take part in fun activities such as sports. It even helps other body systems work properly. But to remain healthy, the skeletal system needs nutrients, exercise, and care.

GLOSSARY

brittle

easily broken

coordination

the ability to move one body part in combination with others

flexible

able to bend

fuse

to bind together with something else

genetics

the study of biological traits inherited from one's parents

splints

medical devices that keep bones straight to prevent pain and promote healing

stem cells

cells that can become one of many other types of cells

symmetrical

having the same shape, size, and appearance on both sides

technique

a method for doing something

SOURCE NOTES

INTRODUCTION: A BAD BREAK

1. Quoted in Mark Long, "Micah Handlogten Will Rejoin Florida for NCAA Tourney after Breaking His Leg in the SEC Title Game," *Associated Press*, March 20, 2024. https://apnews.com.

CHAPTER ONE: WHAT IS THE SKELETAL SYSTEM?

2. Amber Dance, "How Bones Communicate with the Rest of the Body," *Smithsonian Magazine*, March 3, 2022. www.smithsonianmag.com.

3. Steve E. Meadows, "What Are the Most Common Broken Bones?" *South Palm Orthopedics*, November 29, 2022. www.southpalmorthopedics.com.

CHAPTER TWO: HOW DOES THE SKELETAL SYSTEM WORK?

4. Dana Santas, "How to Improve Posture and Relieve Pain with Your Breath," *CNN Health*, June 23, 2021. www.cnn.com.

CHAPTER THREE: WHAT CAN GO WRONG WITH THE SKELETAL SYSTEM?

5. Rob S. Williams, "Fracture Care Treatment: When Is a Cast Not Good Enough?" *Coastal Orthopedics*, December 9, 2019. www.coastalorthoteam.com.

6. Vanessa Caceres, "Exercises for Osteoarthritis," *U.S. News*, March 29, 2024. https://health.usnews.com.

CHAPTER FOUR: WHAT CAN HUMANS DO TO KEEP THE SKELETAL SYSTEM HEALTHY?

7. Heidi Godman, "Essential Nutrients Your Body Needs for Building Bone," *Harvard Health Publishing*, January 1, 2022. www.health.harvard.edu.

8. Becki Andrus, "The Benefits of Warming Up and Stretching," *Orthopedic Associates*, August 15, 2019. https://orthopedicassociates.org.

FOR FURTHER RESEARCH

BOOKS

Samantha S. Bell, *The Human Muscular System*. San Diego, CA: BrightPoint Press, 2025.

Jonas Edwards, *The Skeletal System*. New York: Gareth Stevens, 2022.

Tracy Vonder Brink, *Skeletal System*. New York: Crabtree, 2023.

INTERNET SOURCES

Heidi Godman, "Essential Nutrients Your Body Needs for Building Bone," *Harvard Health Publishing*, January 1, 2022. www.health.harvard.edu.

"Healthy Bones Matter," *National Institute of Arthritis and Musculoskeletal and Skin Diseases*, June 2024. www.niams.nih.gov.

"The Skeletal System," *Innerbody*, July 29, 2020. www.innerbody.com.

WEBSITES

Mayo Clinic

www.mayoclinic.org

The Mayo Clinic provides a variety of health information on its website, including about diseases and conditions of the skeletal system.

MedlinePlus

www.medlineplus.gov

MedlinePlus is an official website of the US government. It provides information on health topics, including the skeletal system.

Nutrition.gov

www.nutrition.gov

Nutrition.gov offers information on healthy eating. It also has articles discussing how diet affects the skeletal system.

INDEX

IMAGE CREDITS

Cover: © Maples Images/Shutterstock Images
5: © wavebreakmedia/Shutterstock Images
7: © Arena Creative/Shutterstock Images
8: © Studio Romantic/Shutterstock Images
11: © New Africa/Shutterstock Images
12: © studiovin/Shutterstock Images
15: © Another77/Shutterstock Images
16: © Simplylove/Shutterstock Images
19: © fongbeerredhot/Shutterstock Images
20: © Halfpoint/Shutterstock Images
23: © Okrasiuk/Shutterstock Images
24: © Jodie Adams/Shutterstock Images
27: © PeopleImages.com-Yuri A./Shutterstock Images
28: © Portrait Image Asia/Shutterstock Images
31: © Leigh Prather/Shutterstock Images
33: © Ground Picture/Shutterstock Images
34: © Inna Kozhina/Shutterstock Images
37: © Mladen Mitrinovic/Shutterstock Images
39: © Real Sports Photos/Shutterstock Images
40: © M. Stocker/Shutterstock Images
43: © Andrey_Popov/Shutterstock Images
45: © Nutnutchar Nav/Shutterstock Images
47: © LightField Studios/Shutterstock Images
49: © nito/Shutterstock Images
51: © Nopphon_1987/Shutterstock Images
52: © MDV Edwards/Shutterstock Images
55: © perfectlab/Shutterstock Images
56: © tonkid/Shutterstock Images

ABOUT THE AUTHOR

Tammy Gagne is an author and editor who specializes in nonfiction. She has written hundreds of books for both young people and adults. She lives in northern New England with her husband, son, and dogs.